上海圣辉制版电脑有限公司 编
刘圣辉 摄影

海峡出版发行集团 | 福建科学技术出版社
THE STRAITS PUBLISHING & DISTRIBUTING GROUP | FUJIAN SCIENCE & TECHNOLOGY PUBLISHING HOUSE

前　言
FOREWORD

随着中国经济的高速发展，国内涌现出了各种设计理念，简直可以说是把欧美的各种设计风潮重演了一遍。稍候国学的兴起，使得国人开始用文化的角度审视周身的事物，随之而起的中式风格设计也为众多的设计师融入其设计理念，可以说21世纪的中国建筑有了中式风格设计复兴的趋势。同时，随着时代的发展，国际设计界也开始越来越重视中式元素和符号的使用。在西方设计界流传着一个观点："没有中国元素，就没有贵气。"中式风格的魅力可见一斑。

中国的古建筑是世界上历史最悠久、体系最完整的建筑体系，从单体建筑到院落组合、城市规划、园林布置等在世界建筑史中都处于领先地位，中国建筑独一无二地体现了"天人合一"的建筑思想。而中国风格的室内装饰，融合了中国字画、瓷塑、明清式家具、中国传统木装修造型及古典型灯饰等，以其综合的艺术色彩，体现出中国的文化艺术氛围。新一代有创意的中国风格的设计师，更是试图从中国几千年文化传统和人文历史中，导引出一种新的具有现代意识的装饰风格。这种风格以静带动、由幽见深、由曲达直地糅合了中国传统思想和现代装饰形式。

中式餐厅与茶楼的装修设计在中式元素上要求十分精细，不仅要体现中国文化，还要与中国人的气节和饮食相关，正确合理运用中式元素才能够真正诠释文化韵味。大多数中式餐厅和茶楼的主人都热爱收藏古董，所以虽然现代餐厅和茶馆的建筑本身不是老房子，却经过精心构筑，非常接近古典风格。在空间比较大的条件下，大多数中式餐厅和茶楼喜欢采用虚实遮挡的重视传统园林建筑手法，空间显得更为静谧和怀旧。明代的花梨木，清代的老红木、榉木，四壁的书画条幅……各陈设相映成趣，浑然一体，古典韵味弥漫其中。

本系列图书以大量的实景案例介绍各种中式风格，这里有传统中式，有现代中式，也有融入中西元素的"上海派"风格……这些经典的中式案例，向正在实践中的设计师和追求中式元素的读者提供了操作范例和素材源泉，欣赏性及参考性兼具。

With the rapid development of China's economy, a great number of design concepts emerge, which has literally replayed the various design trends in Europe and America. After that, the rise of studies of Chinese ancient civilization makes people to look at things around from the cultural aspect. The Chinese style design that followed has been integrated into the design concepts of many architects. The Chinese buildings in 21st century tend to present a renaissance of Chinese style design. Meanwhile, the international design industry pays more and more attention to the use of Chinese elements and symbols. There is a popular idea among Western design field that there is no feeling of noble elegance without Chinese elements, which shows the charm of Chinese style.

Chinese ancient architecture is the most long-standing and integrated building system among the world. From single building, courtyard, city planning to garden arrangement, all are taking the lead of the world architectural history, presenting the unique architectural theory of syncretism between nature and man. While the Chinese style interior decoration blends in Chinese calligraphy, painting, porcelain sculpture, Ming & Qing style furniture, traditional wooden decoration, as well as classical lantern, giving expression to Chinese culture atmosphere through the comprehensive artistic colors. The new generation of creative Chinese style architects even tries to extract a new modern decoration style from the thousands of years' culture heritage and art history. This style mixes together traditional Chinese thinking and modern decoration form in a way of bringing dynamic out of static, deep out of serene and straight out of twist.

In decoration, Chinese restaurants and tea houses are meticulous in the requirements of Chinese elements, which not only represent Chinese culture but also have to be related to Chinese spirit and food. Only appropriate utility of Chinese elements can interpret the cultural appeal. Most owners of the Chinese restaurants and tea house are fond of antique collection, so despite the fact that those buildings are not old houses, they have been elaborately built to resemble ancient building. In a relative big space, most Chinese restaurants and tea houses prefers to adopt the traditional garden design technique of actual and virtual shield that makes the space appear tranquil and nostalgic. The Ming dynasty rosewood, Qing dynasty mahogany, beech and scrolls of calligraphies and paintings on the walls blend into a harmonious whole, exuding a strong classical aroma.

This series of books collect a number of actual projects of various Chinese styles, including traditional Chinese, modern Chinese, and also Shanghai style that combines Chinese and western style. These classical Chinese style projects provide examples and sources with both appreciation and reference for the architects in practice and readers who pursue Chinese elements.

目录 CONTENTS

【蔚景阁】

Wei Jing Ge Restaurant

地址：上海中山东一路2号华尔道夫酒店（近延安东路）

电话：021-63229988

本案中国艺术风格的室内设计与华尔道夫会所的构造交相辉映，营造出奢华的用膳环境。古色古香的用膳环境和精致典雅的中式佳肴，为饕餮食客带来极致的享受。

走进这里，就仿佛进入了一间美轮美奂的阁楼，恰似灯笼的吊灯一个一个从尖顶垂落下来，大幅梅花点缀餐厅主背景墙，暗香涌动，静谧而充满格调。开放式的椽木天花板以及传统的中国栅格结构增添了『阁楼』的气氛。中式的花格罩门，配以深色纱帘，似隔非隔，营造出一个个淡雅空间；青砖砌墙配以石雕版画，处处彰显着与众不同的气质。有别于一般的中餐厅，本案在继承中式博大精深的同时，更发挥现代中餐的创意，餐厅配置了几间个性化设计、布置典雅的包厢，贯穿雅间的中央走廊设有现代的背光灯墙，地面铺以宝蓝色地毯，韵味十足。中式的多宝格展示着中国珍宝和各式有趣的物品，让置身其中的顾客时刻感受着古典与现代融合的魅力，仿如穿梭亘古。本案在细节处也精致至极，每一处灯光的点缀，每一处色彩的铺贴，每一处材质的运用，都使本案鹤立鸡群。

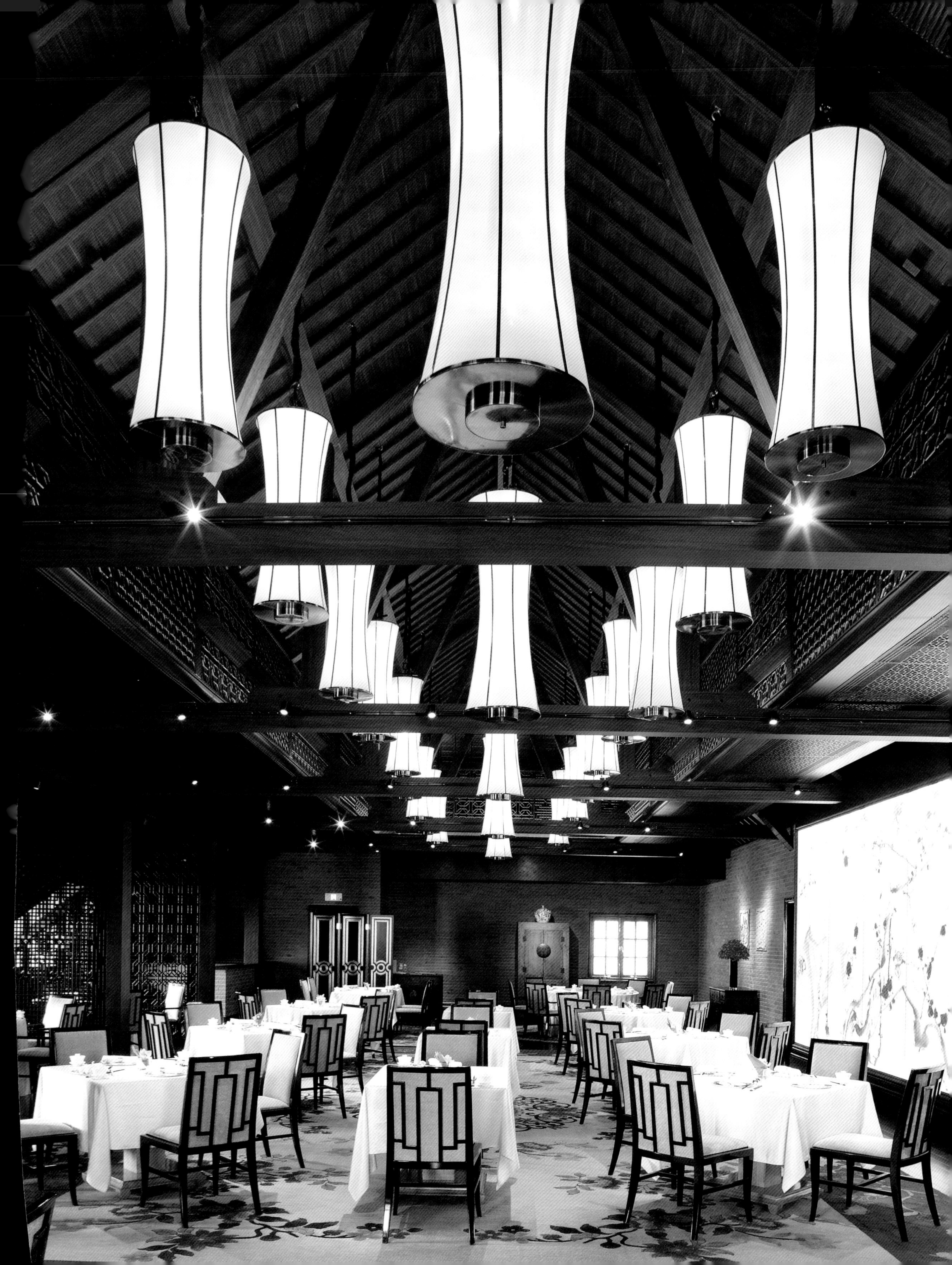

The Chinoiserie interior design of this project complements the Waldorf Astoria Club to create a luxury dining environment. The antique dinning space together with delicate Chinese dishes brings guests a perfect Chinese gourmet.

The moment they walk in, guests are entering a splendid attic, with pendant lamps like lanterns hanging from the spire against the plum blossom thematic wall which suggests a felling of hidden fragrance, presenting tranquility and style ambience. Open rafter wood ceilings enhance the 'attic' ambiance which is also accented with iconic Chinese lattice work. Chinese lattice barn doors are decorated by dark color sheer curtain to build a quietly elegant space. Black brick wall are matched with stone carving prints to demonstrate a distinctive temperament. Unlike ordinary Chinese restaurant, while inheriting profound tradition Chinese style, the project further exerts modern Chinese creativity. The restaurant has several characteristic design private rooms with elegant decorations. The central corridor through the private rooms features contemporary, back lit walls and royal blue carpet. The Chinese curio cabinet that showcase Chinese 'treasures' and interesting finds leads guests to feel the charm of blending classical and modern. The details are also refined to the best, every single lighting, color and material is designed to make this space stand out in the crowd.

【鸟语花香】

Birds' Twitter and Fragrance of Flowers Restaurant

地址：杭州保俶路33号

电话：0571-85160077

『鸟语花香』用在这个餐饮空间里十分贴切。严格地说设计师在着手设计的时候并未严格遵守传统的中式风格，并将其贯穿其间。设计师在设计思路的整合上巧妙地取悦于国人向往的『鸟语花香』，并将其作为贯穿空间的设计主线。在表现这个题材的时候，设计师采用了写实的方式，比如，你可以在通道的墙面装饰上看到带有该主题性质的装饰图案，有花有鸟有文竹……通过这种写实的方式去表现自己要表达的设计主题。为了强调中式的氛围，设计师在利用横向线条装饰的间隙，间插进去一些中式书法图案，以一种隐约的方式来告诉我们这是一个需要表达出中式风格的设计。为了去除五千年文化的沉重，设计师在色彩的表现上没有延续中式的厚重，取而代之的是白色的清爽。设计师意在营造一个既带有中式文化氛围，又比较轻松的用餐环境。在玩味中式元素的基础上，设计师如顽童一般并未满足，他在公共区间局部的吊顶处理上，大胆地采用了一些欧式的元素。华丽的欧式色彩埋藏在总体的中式氛围中，以一种夺人眼球的方式，达到了一个奇怪的表现方式。这种冲突在大量的白色环境中偃旗息鼓，让『鸟语花香』跳脱了字眼上的活跃，让空间充满玩味。

The name Birds' Twitter and Fragrance of Flowers fits this dinning space perfectly. Technically, the architect did not strictly follow traditional Chinese style while designing this space. He adopts the life style Chinese people long for as its motif throughout the whole space, namely birds' twitter and fragrance of flowers.And realism is used to present this motif, e.g. on the walls at the corridor, there are decorative patterns with different features, including flowers, birds, bamboos and etc to present this motif. To emphasize Chinese style atmosphere here, the architect occasionally inserts Chinese calligraphy patterns among the horizontal line decorations, subtly telling us that this is a Chinese style decoration. In order to eliminate the heaviness of 5000 year old culture, the space did not stick to Chinese colors; instead, it uses refreshing white. The restaurant intends to exhibit a dinning space with Chinese cultural temperament and yet remains relaxing. In addition to the Chinese elements, the architect playfully and boldly uses European elements in part of the ceilings in public area. Splendid European colors hidden in the whole Chinese atmosphere has found an odd expression in an eye catching way. This conflict ceases in the huge white environment, to free the name Birds' Twitter and Fragrance of Flowers from literal characters to active image, leaving guests pondering in this space.

【和平官邸】

Peace Mansion

地址：上海市汾阳路158号

电话：021-64375193

私房菜的风行推动了精明的商家在餐饮经营的选址上注重标新立异。空间的独立性、私密性是此类私房菜在经营选址时的首要条件。静谧的私家小院被商家稍加打扮，瞬间成为了独立经营私家菜的承载舞台。设计师根据商家经营菜品的特点，果断地将空间定性为新中式的装饰风格。无处不在的混搭杂糅在这个颇具历史韵味的欧式小洋楼里。中西合并的庭院，中式的木屏风被刷上了跳跃而艳丽的洋红色，与周边清透的玻璃顶及朴实的建筑外观形成了一种反差强烈的动静对比冲击。室内空间则被设计师隔成了一个个独立的包间。保留之前的欧式装饰硬风格，设计师巧妙地通过软装饰品的点缀作用，杂糅进她所想要的中式元素。于是，你可以在角落里看见充满唐风古韵的装饰小摆件，甚至在一个看似无用的角落里发现一个古旧的中式柜子。走走逛逛，你会发现在小洋楼里面绝对是欧式包围中式，而在小洋楼的外面却是中式包围欧式。这种交错的混搭方式，让每一处、每一点都能为不同喜好的客户提供小小的惊喜，让每个用餐的客人有不同的选择，在惊喜和欢乐中愉快进餐。

The popularity of private kitchens has driven smart merchants to choose unconventional location for their restaurant. Independence and privacy are the primary elements for private kitchens. The serene private courtyard has been turned into an independent private kitchen by a little decoration. Based on the feature of the cuisine here, the architect deliberately defined its neo-Chinese style for this space. Mix match could be found everywhere in this historical European style building. Upon arrival, a courtyard combining Chinese and Western architectural styles comes into sight. The Chinese wooden screen is painted to the dynamic color of magenta to create a strong visual contrast against the clear glass ceiling and simple outward appearance. The interior is partitioned into independent private rooms. While preserving its original European style hard decoration, the architect tactfully blends Chinese flavor into this space with soft adornments. You will run across in an obscure corner, small decorative items with ancient charm, or even an archaic Chinese style cabinet. Walking around, you will find European style is surrounding Chinese style in the interior, while it is the other way round outdoors. With scenes change at every step, the restaurant gives small surprises at every point and offers various choices for guests with different preferences, creating a joyful and pleasant dining environment.

渝 宴

Yu Yuan Restaurant

地址：上海四川北路1318号盛邦国际4楼（近武进路）

电话：021-3635 6013

如何打破川菜的低消费形象，设计师提交了一份自己的理解和答案。本案并没有延续传统川菜馆的乡土装饰风格，而是通过时尚的设计概念转变川菜的乡土气息。花为设计主题，也是设计师的设计主线。在空间的平面布局上，回转曲折的弧线如飘带般带来风的气息。吊顶的处理上，设计师从叶片中获取设计灵感，用格栅的形式来表现叶脉，通过由上及下的飘带式的隔断串联起顶面与墙面。顶面是叶片，那么墙面就是花朵和花瓣了。通过手绘的花朵图案，设计师为不同空间赋予不同类型的花，有牡丹，有玫瑰，串联起来，设计师的设计意图昭然若揭——大风吹起之下，花叶漫天飞。为了体现这一自由的花香意向，设计师在局部墙面的灯光处理上选择了LED灯，在变幻的色彩中带来迷幻的意境。川菜已不仅仅是传统意义上的川菜了，在设计师的手中，已经把它的泼辣做派随风般开枝散叶，赋予了新含义的川菜悄然间也散发出时尚的味道。

WOMEN

How to break the low-end image of Sichuan restaurant? The architect presents here his own answer to the question. The restaurant did not continue local decoration style of traditional Sichuan restaurant; instead he replaces the local flavor with stylish design concept. Flower is the motif throughout this space. Swerving and winding curves in plan arrangement bring the smell of wind. The ceiling was designed as a leaf, by using grates as leaf veins. Ribbon partitions from top to bottom links the ceiling and the wall which is decorated by flower and petal patterns. By using hand drawing floral patterns, the architect decorates different spaces with different flowers, such as peony and rose,. The intention of this design is obvious – imagine the flowers and leaves flying all over the place when the strong wind comes. The architect chooses LED light in some wall finishings to create an enchanting ambience. Sichuan cuisine is more than a traditional cuisine here, because its hot style is all spreading with the wind. Endowed with new implication, Sichuan cuisine is silently showing its potential to be served with a chic flavor.

16
小心玻璃

小心玻璃

【中国宫殿】

China Palace Restaurant

地址：上海太仓路181弄新天地北里17号楼

电话：021-53069988

中国宫殿的符号是怎样的？如何去体现中国宫殿的意境呢？本案设计师是这样来处理的：首先截取中国宫殿中最为出名的皇家园林作为表现符号，于是你能看到很多皇家园林的设计手法呈现其间。月影门、花瓶洞、葫芦孔、亭台楼榭……作为设计符号散落在空间四处。为了体现宫殿的厚重，设计师在局部立面处理上使用了石材。为了避免石材在有限的室内空间里显得过于厚重，在灯光的处理上注意阴影的处理，使得石材若隐若现，弱化石材的硬与重的感觉。在软装饰品的设计上，设计师巧妙地以清宫廷里大臣的官帽作为灯罩造型。或写意般正向吊装，或写意般反向吊装，在正正反反间，在阴阴暗暗里，将宫殿的讳莫如深和一入宫门深似海的感觉表现得淋漓尽致。整体空间设计饱满庄重，充分体现出经营者需要的空间氛围，达到了商业设计的经营要求。

What is the symbol of palace? How to reflect the atmosphere of palace? The architect of this restaurant has solved these problems in his way. First he picks the most famous palace, royal garden, as its symbol. Hence you can find many design approaches adopted from royal garden, such as moonlight shadow gate, vase shape doorway, gourd hole, pavilions and etc, scattering all around this place. Stone material is applied in the finishing of vertical façade, to express the stableness and solemnness of palace. Yet to prevent stones from appearing too thick in limited space, shadows are added in lighting effect to weaken the hard and heavy feeling of stone. In soft decors, the architect tactfully uses officials' hat of Qing Dynasty as lampshades and installed them upright or inversely, vividly revealing the reticent feeling of palace. Being full and magnificent, this space well creates the requested ambient and meets the demands of business operation.

【燊公馆】

Shen Mansion

地址：上海浦东新区世博大道450号（近雪野路）

电话：021-22060488

一提到公馆，就给人高厅朗阁、雍容华贵的感觉。进入大厅，一个巨大的牌坊式的入口映入眼帘，其占据了两层楼高的玄关（姑且视为玄关的功能）带来了巨大的气势，一幅『逍遥杖国』的牌匾指引着用餐的人前往二层的包厢消费。在首层的大厅，设计师为了体现大气如官堂的效果，装饰处理上皆采取大块面的方式，整体的设计拒绝小气。这样的处理手法，给人造成一种高档消费的心理暗示。拾阶而上，二层的设计延续着大气二字，奢侈的空间处理，不论在走道上还是在包房里都让位给高级艺术品的展示。虽然在用材和灯光的色调处理上，尽量体现出温润婉约，低调奢华的东方气质，但是仍然掩饰不住设计师想要表达的经典复古和雍容华贵，将上海滩的流金岁月精致呈现。整体的空间注重人文关怀，点造型的处理上复杂，线造型的处理上疏朗，面造型的处理上则强调大气统一，将美食融于文化，用文化为美食推波助澜。

安全出口
EXIT

The word "mansion" always reminds people of a dignified and graceful house with high walls and long hall. Upon arrival at the restaurant, a huge archway that functions as the entrance of the hall comes into sight. After that is a magnificent two-storey foyer, in which a plaque leads the way to private rooms on second floor. In order to create grandness at the lobby on first floor, the architect uses only large materials in decoration, which extends to guests implications of upscale consuming. The decoration of second floor continues to adopt this character. The spaces, both at the aisle and in the private rooms are used to display refined artworks. Although the materials and lightings are designed to create a mild, sumptuous traditional oriental temperament, this cannot cover the classic, elegant charm as well as precious times of the past that the architect tries to present. The entire space attaches importance to cultural concern and is designed with complex point forming, lichtung line shaping and grand integrated plane figuring. Here gourmet is integrated in culture that helps promote the gourmet.

安全出口
EXIT

風和
日暖

【黔香阁】

Guizhou Pavilion Restaurant

地址：上海虹中路525号

电话：021-6401977

黔香阁里的一个『黔』字，透露了这是一个经营贵州地区菜系的餐饮空间。既然是经营贵州菜系，那么在餐饮空间的内装设计上就必须符合贵州地区的地域特色，就必须营造出具有贵州地方风情和民俗文化的氛围。在餐厅休息厅处，那匠心独运的砖雕木刻，那古色古香的黛砖粉墙，小到充满灵气的一柱一石、一孔一景、一窗一棂，无不显示出被古人称作『老远的贵州』那丰厚的文化底蕴，无不透示出风光无限的贵州那雄山秀水的神韵。餐厅的装修浑然大气，充分利用空间的高度来表现贵州地区大山的气势。用色上以灰、白为主调，质朴的装饰材料用于表现空间的淳朴。为了体现贵系菜肴辣的特色，设计师在局部用色上选择了鲜艳的红色。在整体灰白色调中红色无疑起到了刺激感官的作用。整体的设计古朴典雅，体现了业主所致力发展的文化餐饮概念，符合经营需求。

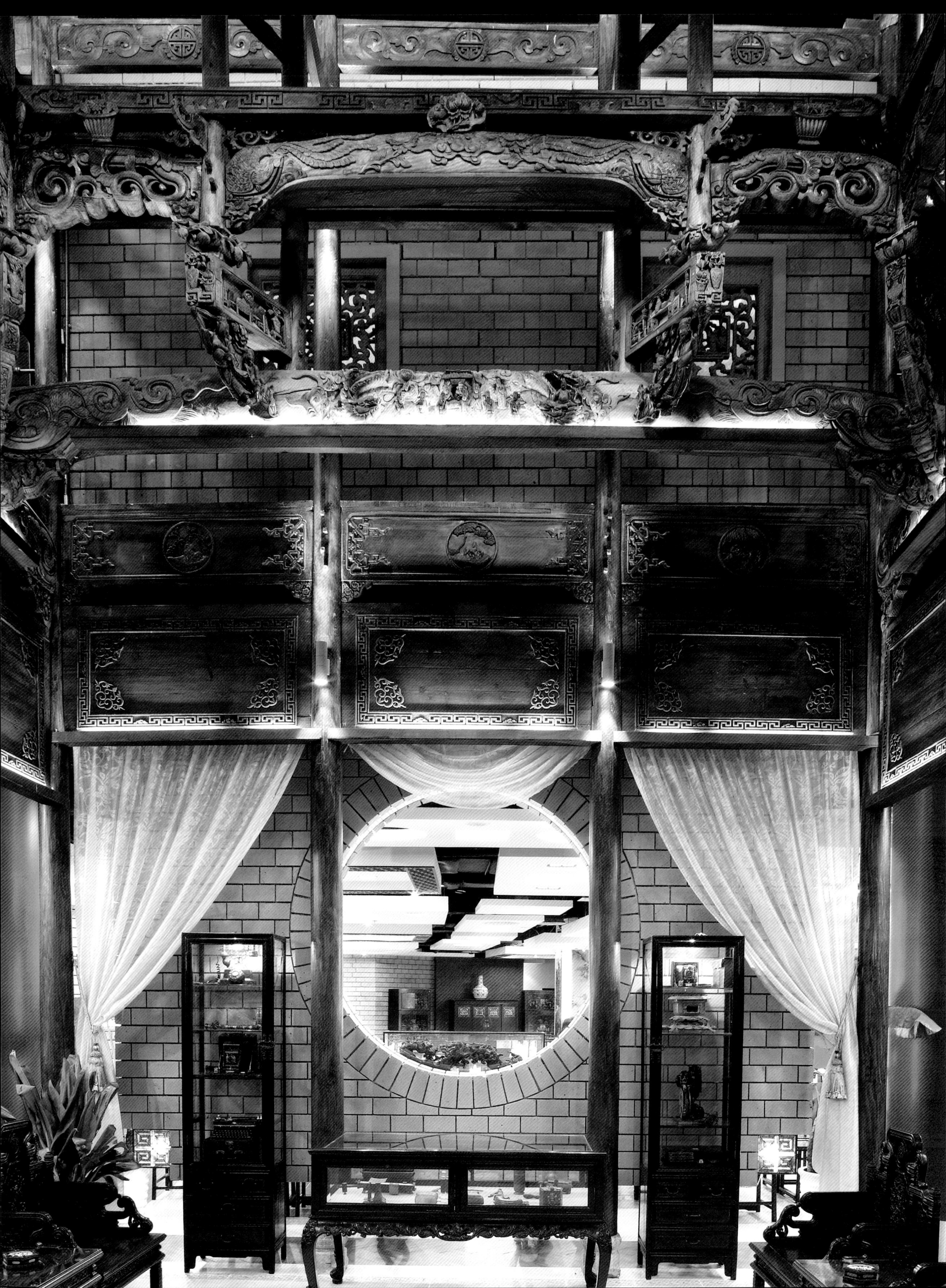

The name Guizhou Pavilion Restaurant suggests it is a dinning space features Guizhou cuisine. Therefore the interior decoration is designed to build an environment with Guizhou flavor and folk culture.

Upon arrival, people will be attracted by the unique tile carvings, wood engravings and archaic brick walls in the lounge hall. Every piece of the columns, stones, holes, windows and latticework reveals the profound culture heritage and infinite scenery of the so-called "remote Guizhou" in ancient time. The architect fully utilized the height in this space to represent the vigor of Mount Guixiang in Guizhou, creating a majestic and natural environment. To give expression to the characteristic "spicy" of Guzhou cuisine, the architect adds bright red in coloring of partial space. Against the gray and white background throughout the whole space, red has undoubtedly stimulated the visual sense. The overall design is simple and elegant, which demonstrates the owner's business concept of "integrating culture in catering".

八十七神仙卷

【凌泷阁】

Ling Long Ge Restaurant

地址：上海虹许路951号2楼（延安西路口）
电话：021-32070077

凌泷阁取『玲珑』谐音。玲珑，有精巧细致之意，古人也用玲珑描述玉石碰撞的清悦之声。本案以凌泷命名，恰如其分地将精巧细致的设计手法和仿佛铿然有声的空间概括和提炼。

设计师力图通过色彩，来奠定空间浓厚的中式韵味和自己别出心裁的设计。要用颜色表现中式风格，最鲜明的莫过于中国红，于是便有了大面积鲜红色的吊顶和墙面，这为整个空间的中式风格打下了基础。这种红色仅限于在门厅和过道等相对公共的空间使用，到了用餐的包厢内，则很少再见到这样鲜明的红色，而代之以白色、粉黄等较为淡雅的颜色，这其实是设计师结合色彩心理学的设计。门厅和过道等公共空间大量的红色使消费者一进门就受到强烈的感官刺激，引发消费者的食欲。包厢内的用色则可以稍微平复消费者的心理，使消费者开始静下心来细细品味每一道佳肴。设计师自身的设计理念和空间本身的属性实现了完美结合。玲—麟，珑—龙，中国古代以其象征祥瑞，夏商时期的食具器皿上也多用饕餮纹装饰，代表丰盛的食物，空间中大量珍兽石雕的陈设再一次强化了中式的主题和空间的属性。

凌瓏閣
凌瓏閣

安全出口
EXIT

Ling Long means exquisite and delicate, the ancients also use this word to describe the tinkling sound of jade colliding with each other. Named as Li Long, this project appropriately summarizes and extracts exquisite design techniques to refine this seemingly loud and clear space.

The architect intends to establish a strong Chinese flavor and ingenuity in this space by use of color. No other color expresses Chinese flavor better than Chinese red. Thus this space is designed with extensive red ceiling and wall surface, which lays the basis of Chinese flavor. Yet this color is only applied in public area such as hallway and corridor, the private rooms are instead, decorated in light colors such as white and pastel yellow, which is deliberately arranged according to color psychology. The vast red color in public area extends a strong sensory experience and stimulates every cell of the taste bud to stir one's appetite. Nevertheless, the colors in the private room are designed to slightly pacify one's mood and direct guests to savor every delicacy. The design concept of the architect has been perfectly combined with the nature of this space. In Chinese, the name ling and long are respectively homophones for unicorn and dragon, both of which are ancient legendary creatures. During Xia and Shang Dynasty, Taotie (a mythical animal) pattern also appears quite a lot in tableware and utensils to represent abundant food. The plenty precious creature stone carvings displayed again underline the Chinese motif and nature of this space.

馬蹈

【晶彩轩】

Jingcai Pavilion

地址：上海虹桥路1937号

电话：021-62709577

设计师在此间餐厅表现中式风格的时候，强调的是中式元素的点缀性应用。入口的琉璃装饰片附挂在灰色的整体装饰墙上，而在建筑梁的位置上为了遮羞，设计师选用了带有故事性图案的砖雕加以处理。入得厅堂，随处可见的中式木雕呈现在视线所及的立面、顶面上。虽然图案繁杂，但是设计师在使用的时候注意色彩的搭配，反而弱化了繁杂的中式图案给消费者带来的焦躁心理。比如，在顶面的处理上，设计师为了弱化繁杂的中式图案，特意选择了白色油漆进行处理，在立面的处理上，设计师则利用黑色的石材及深色的砖来搭配复杂图案的中式柜子。深色带来的稳重感将红色的木雕部分的浮躁压制下来，从而让空间既保留了中式高贵的质感，又符合引导消费者理性消费的商业需求。整体的空间落落大方，不浮躁，保留了中式传统的原汁原味。

The architect tries to demonstrate Chinese style in this restaurant by emphasizing the decorative function of Chinese elements. At the entrance, lazurite overlays are attached on the gray wall finishing. In order to hide the beam, tile carvings that have images telling stories on them are placed on it. In the hall, Chinese wooden carvings are used on all the vertical and top facades. Although the pattern is complicated, but the color arrangement has weakened the restless feeling that complicated Chinese patterns bring. For example, white painting is used particularly on top facades to reduce the complication of Chinese style pattern; while on vertical facades, the architect adopts black stones and dark bricks to match the complicated pattern of Chinese style cabinet. The stable sense that dark color brings suppresses the fickleness from wooden carving, preserving Chinese nobility in this space and yet complies with the business requirement to guide rational consumptions. On the whole, the space is natural and graceful that maintains original flavor of traditional Chinese style architecture.

拾贰

拾贰

【屋里香】

Wulixiang Restaurant

地址：上海南昌路164号（近思南路）

电话：021-53065462

这个餐厅像是艺术家的作品展示舞台，整体的设计更像是草根文化的整合。在硬装的设计上设计师没有花费太多的精力，简单到连墙面都直接刷白处理，用一种很直接的方式来展现砖墙的本色质感。与之相反的是设计师花费了很多精力在软装陈设上。之所以说是草根文化的整合，是因为餐桌、椅子及其他经营餐馆需要的硬件家具，怎么看都像是设计师特意从民间收集来的旧家具。为了彻底体现草根精神，设计师还把草编的篮筐一束一束地捆在一起，挂在高处作为装饰，把农闲时分取乐的戏曲道具及戏曲剧目，或挂在高处或摆放角落。若是伴上点依依呀呀的戏曲唱腔，那么一个热闹的旧式小酒馆就生灵活现地开张了。设计师还戏谑般地把欧式吊灯高高地悬挂起来，洋气了一把。于是在艳丽的色彩下，在轻松的氛围中，好酒好菜伺候着，『屋里香』就这样开张了。这样的设计明确了经营者的经营方向——为人民服务。

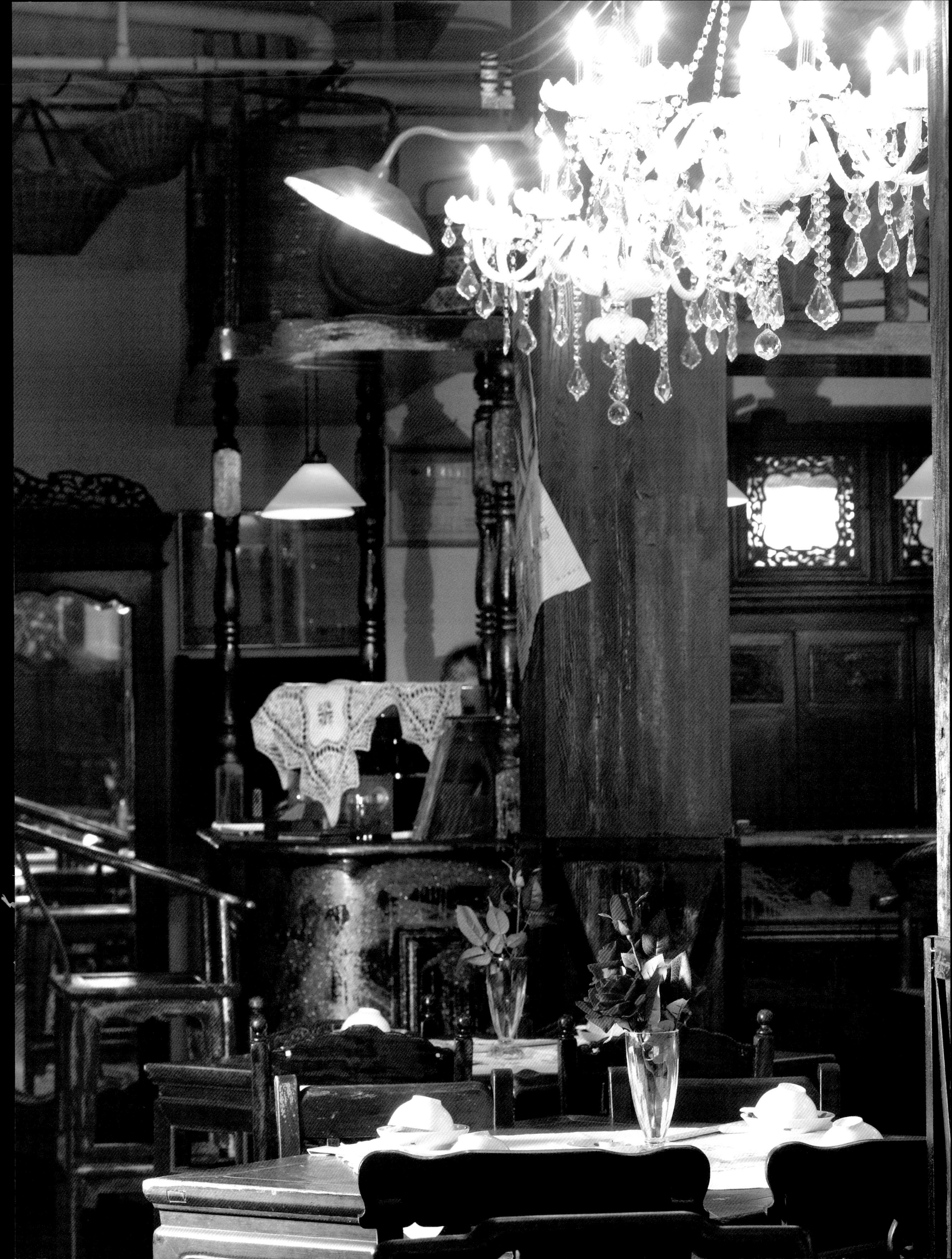

The restaurant is more like a stage for artworks as its entire design feels like an integration of grass roots culture. The architect did not spend much effort on hard decoration. The walls are simply painted white to display straightforwardly the natural texture of brick walls. On the contrary, the architect lays emphasis on soft decoration and furnishing. The reason why this space is an integration of grass roots culture is because the tables, chairs and other necessary furniture for a restaurant here appears to be old furniture collected from the folk. The architect intends to fully express grass roots culture in this space, by hanging bundled grass basketries overhead and lay Chinese opera props and plays on high or in the corner. With a few Chinese opera singing, an old style boite would have lively come into being. To show the resolution for a serious business, the architect playfully hangs European chandelier on the ceiling, adding Western flavor to this space. Thus, among beautiful colors, relaxing ambient and delicious food, Wulixiang opens with a decoration that defines its business direction: serve the people.

甘露寺

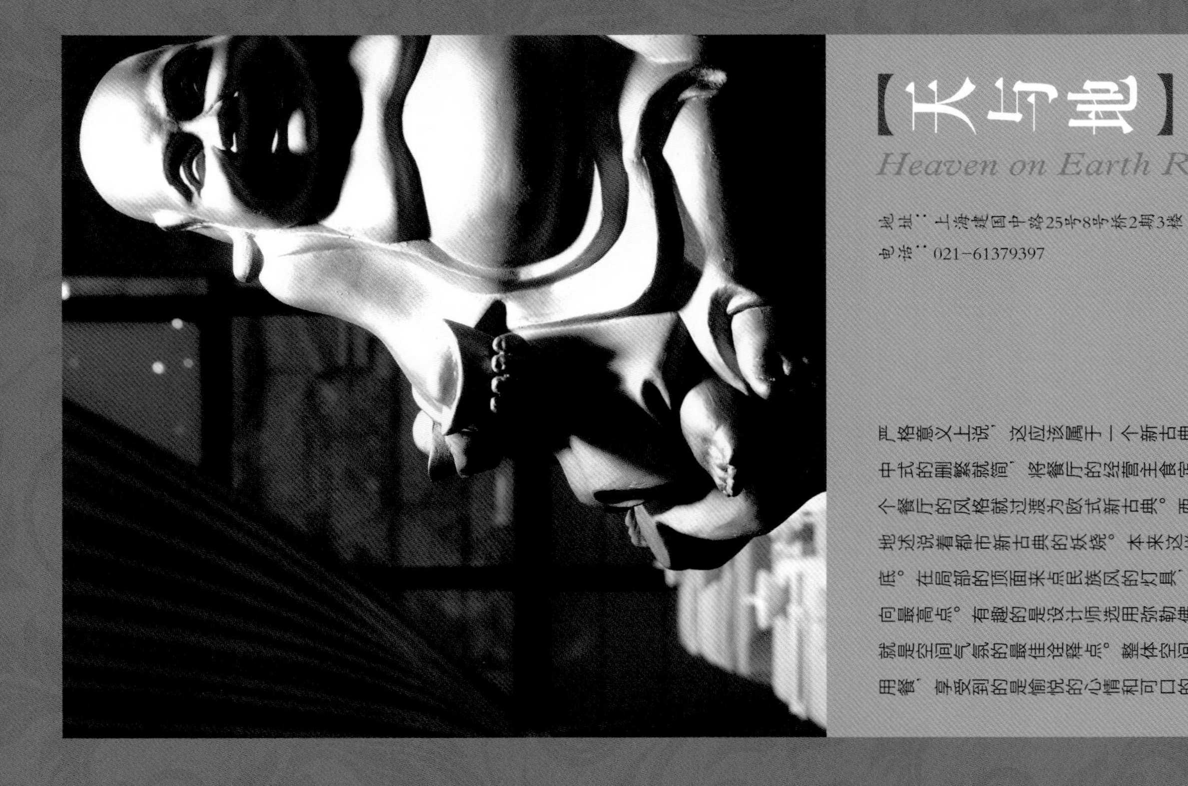

【天与地】

Heaven on Earth Restaurant

地址：上海建国中路25号8号桥2期3楼
电话：021–61379397

严格意义上说，这应该属于一个新古典混搭的餐厅。餐厅的门厅是不折不扣的新中式风格，通过对古典中式的删繁就简，将餐厅的经营主食定位为中餐。通过隔断博古架的过渡，在成排摆放的弥勒佛中，整个餐厅的风格就过渡为欧式新古典。西式的餐桌摆放方式，华丽丽的欧式吊灯，艳丽的色彩，不慌不忙地述说着都市新古典的妖娆。本来这样中西合壁的混搭方式已经很热闹了，可设计师却将热闹进行到底。在局部的顶面来点民族风的灯具，艳丽的色彩加上层次分明的灯具组合，彻底地将空间里的气氛推向最高点。有趣的是设计师选用弥勒佛作为装饰点缀，于是你会看到博古架上弥勒佛大笑的表情，简直就是空间气氛的最佳诠释点。整体空间虽然杂但却不乱，混搭下的各个元素运用得恰到好处。坐在里面用餐，享受到的是愉悦的心情和可口的美食。

消费满100元
20元

天与地
中华料理
午市商务套餐
单人套餐15元
四位，六位，八位
168-368元
歡迎光臨
Welcome

安全出口
EXIT

Technically, this restaurant is designed with neoclassical style. The entrance hall is modern Chinese by simplifying classical Chinese style. The restaurant features Chinese cuisine. An antique shelf filled up with rows of Maitreya Buddha statues Partitioned the restaurant, and s as a transit to the European neoclassical style decoration behind it. Western laying of dinning tables, gorgeous chandelier and bright colors deliberately tells the enchanting urban neoclassicism. This blending of Chinese and Western style is supposed to be boisterous enough; however, the architect decides to carry this through to the whole space. Some of the top surfaces are decorated with ethnic lighting, which combines gay colors with well-bedded lighting and brings the atmosphere to a climax. The architect interestingly uses Maitreya Buddha as an ornament. Therefore the laughter face of Maitreya Buddha best illustrates the atmosphere of this space. The entire space is complex yet well organized, with every element in this blending properly used. Served with tasty food, guests can expect a very delighted dinning experience here.

【荷餐厅】

Lotus Restaurant Restaurant

地址：上海长宁路641号（近安西路）
电话：021-52382919

出淤泥而不染，濯清涟而不妖。『荷』自古以来就是文人墨客竞相称颂的一个意象。发展至今日，在设计领域她依然备受设计师的青睐。本案正是一个以『荷』为主题的现代餐厅设计。

在引用『荷』这一意象的时候，设计师的高明之处在于并非直接地硬搬照抄，而是运用重复、解构、抽象等美学原则，融合现代主义的表现手法，将『荷』渗透到空间的每个层面，实现主题与表现完美的结合。比如装饰画的使用，既有偏写实的也有经提炼抽象的，既有解构的组画也有整幅的大图和整面的墙绘。通过各种不同手法的运用，将主题元素灵活地融入到空间的各个角落。

天花用紫黑色的亚光墙漆，仿佛是静谧的荷塘。采用高技派手法直接外露的中央空调和通风管道直接涂成紫黑色，恰似水面下若隐若现的枝叶。地面同样是紫黑色的亚光仿古地砖，天花的元素若有若无的倒映，又形成一道风景。为了调和深色天花和地面的压抑感，中间穿插了大量的浅色竖向线条元素，比如自然弯曲的线条隔断和垂直的吊灯，这又恰似出水的茎枝和莲蓬的形态。整个空间浑然一体，给顾客带来超越就餐本身的体验服务。

海最能吃辣的人

“She comes from the ooze, yet remains stainless. Though baptized by ripples, she stands modest.” Since ancient times, lotus has been an image praised by various poets, and till today she is still long favored by architects in design field. This project is just a modern restaurant with lotus as the design motif.

While introducing this image, the architect did not directly copy everything mechanically, instead, he brilliantly uses aesthetic principles including repetition, deconstruction and abstract to blend in modernistic technique of expression and penetrates “lotus” into every layer of the space to achieve perfect combination of the theme and the expression. For example in the use of decorative paintings, there are realistic and abstract style, deconstruction paintings in series and also full wall painting. By adopting various techniques, the architect subtly integrate the theme into every corner of the space.

The ceiling finished by purple black matte painting looks like a tranquil lotus pond. High-tech techniques has painted the exposed central air-conditioner and ventilating duct into purple black, that appears to be indistinct branches and leaves under water. The floor is also finished by purple black matte archaize tiles, with elements of the ceiling reflected on it, forming another scenery. In order to offset the sense of constraints brought by dark color ceiling and floor surface, plentiful light color vertical lines are interspersed in them, such as naturally curved line partitions and vertical pendant lamps which exactly resemble the shape of the stem and lotus receptacle. The space is an integral whole that arouses confusion whether you are above the lotus pond or under water, bringing to guests a relaxing experience more than dinning itself.

【五观堂素食】

Wuguantang Vegetarian Restaurant

地址：上海新华路349号（近定西路）
电话：021－62813695

佛说『一花一世界，一叶一如来』。一说到素食，在装饰设计上往往趋向简约风格，强调的是空间的意境和禅意的营造。首先来看主入口处的建筑外观，大量白色面处理，营造一种安静的气氛，古铜色的店招和古铜色的门套框，在形成强烈对比的同时更是加重了安静气氛的营造。入得厅堂，设计师也是简简单单不留痕迹地进行设计。黑、白、灰是空间的主色调，偶尔的洋红色也只为了撩拨空间的静谧。毕竟是营业场所，太过素雅也不符合商业规律。设计师用些轻飘的纱幔、艳丽的骨牌，还有古朴的中式博古架来分隔空间，让空间隔而不断。在保证整体氛围一致的前提下，各个空间又各有风味，让空间在热闹中淡定，在淡定中富有品位。包房里，设计师注重字画的装饰作用。在字画图案的选择上多以荷花作为主要装饰元素，取莲出淤泥而不染的意境及佛教中莲花的象征性意义，来表达对禅的敬畏。整体空间氛围营造到位，点到为止，在满足商业需求的前提下，赋予空间以禅文化的含义。

五觀堂素食
五觀堂素食
五觀堂素食
349
五觀堂素食
五觀堂素

Buddha says: To see a world in a flower and a heaven in a leaf." Speaking of vegetarian restaurant, simplicity is the basic element in decoration to lay stress on atmosphere and Zen. The façade of the main entrance is covered by white color, creating a tranquil atmosphere. Also, no trace of deliberate design is found in the simple decoration of the hall. Black, white and gray comprise the dominant colors of this space, with occasional magenta teasing the tranquility. However, too much simplicity is not appropriate for such a commercial place. Therefore, the architect uses light gauze curtain, colorful dominoes and unsophisticated Chinese antique shelf to partition the space, yet not to cut off the spaces. Each space has its own feature without breaking the consistent ambience in the whole space. The place remains calm in the hustle, and tasteful in the calmness. In decoration of private rooms, the architect mainly uses calligraphy and paintings, among which lotus is the most commonly seen element to symbolize purity and innocence as well as to pay respect to Zen in Buddhism. The entire design creates just the right atmosphere that not only satisfies the business requirement, but also endows the space with Zen culture.

禅心

悦心

菜根譚
唐詩宋詞元曲
四書五經

【颖尚精菜坊】

Ying Shang Restaurant

地址：上海虹秀路51号（虹泉路西）

电话：021-64801717

走进颖尚精菜坊，仿佛置身于花的海洋，古钱币造型的拱门，树木与假山环绕，不禁让人想起张旭《桃花溪》里的『桃花尽日随流水，洞在清溪何处边？』

通透的玻璃房设计，在阳光充足的空间中，一切物体颇具独特的表现力。几处桃花盛开时，小桥淌水、树枝蟠曲，配以假山环绕，仙雾弥漫，更有仙境般的视觉感受，身临其境感受用餐的愉悦心情。往里渐进，拱形的紫藤花架分隔出就餐区域，一盏欧式吊灯的点缀，营造出一种温馨而浪漫的气息，特别有小花园的意境。几座小木屋置于其中，石墙与青瓦颇具古朴特色，麻绳收边、原木柱子、圆形窗格，更有屋檐下的竹编小夜灯，凸显设计师营造氛围的用心良苦。木屋室内色调呈重木色，却因陈设和装饰的不同而各有千秋，有的以野生麋鹿的头部标本装饰，气派凝重；有的布满白色樱花，清新自然；还有的大幅手绘临摹韩熙载《夜宴图》，色彩绚丽，凸显其古典韵味。整个餐厅设计从细节中体现菜坊的品牌理念，环境受之，心灵感之，细细品之，足矣。

古韵

06

Upon entering the ancient coin shape arc door, one seems to have stepped into an ocean of bloom. Surrounded by trees and rockery, one cannot help but thinking of the lines in the ancient poem Peach Blossom River: "Peach petals flow with waters all the day; Where the cave he beside the clean stream may?"

The whole space is a transparent glass house with sufficient sunlight, where every object has its own distinct character. Peach blossom, bridge above streaming water, bent and winding branches are encircled by rockery against the misty background to paint a fairyland vision, bringing a pleasant dinning experience. Advancing gradually, the archy wisteria pergola partitioned the dinning area dotted by a European pendant lamp, creating a cozy and poetical atmosphere like a garden. A few cabins are disposed in this area, with unsophisticated stone wall, gray tiles, hemp rope, log pillar, circular window pane, and even little bamboo weaving night lamps under roof that reveals the architect's painstakingly approach to set the mood of the space. The tone in the wood cabins is dark wood, and yet their feeling varies from each other because of the furnishings and decoration inside. For example, the cabin decorated with specimen of head of wild elk appears to be regal and dignified; while the one covered by white sakura blossom looks pure and fresh; also, there is a tracing painting of Banquet at Han Xizai's Residence in one of the cabin, to highlight the classical appeal. The entire design of the restaurant fully reflects its brand concept in all details, gradually transferring the idea from the environment to one's heart.

花草丛中笑
园外观其貌

【新农庄】

New Farm Restaurant

地址：上海剑河路2396号（近延安路高架）

电话：021－62422277

新农庄以四合院落为格局，飞檐翘角的古建筑轻巧玲珑，小桥、假山、翠竹、桃树凸显园林风格中的『崇尚自然』的主张，建筑与环境有机地融合为一体。

院落中弥漫着静谧而古朴的中国风，木质的休闲桌椅与遮阳伞流露出轻松惬意的气息。依院落中的长廊渐进，农家风情的主题跃然纸上，农家拉货车、高脚竹编柜子、木桌木椅、青瓦砖墙处处喻意浓浓。餐厅空间中叶形木质镂空隔断及竹帘，矩形及多边形的窗户栅格造型，将空间分隔得井然有序。石雕望柱与竹制栏杆的结合极为别致，中空的设计又延续了上下层的空间。红色花轿喜庆热烈，充分体现中国传统风俗的人文特色，随处可见的石雕、砖雕、木雕为空间增添了古典韵味。几株盛开的桃花点缀在室内，更是『人面桃花相映红』，追求宁静自然、朴实无华的清新风格，创造出与自然环境协调共生、天人合一的艺术综合体。

28

12
18
11

Located in a quadrangle courtyard with overhanging eaves, New Farm is designed with a light and exquisite architectural appearance. The bridge, rockery, bamboo and peach trees in it highlight the proposition of "advocating nature" in garden style, which organically integrated the construction with the environment.

The courtyard is suffused with a tranquil and simple Chinese flavor. The wooden leisure tables and chairs exude a relaxing and cozy ambient. Along the corridor in the yard, the motif of country lifestyle are vividly revealed by the cart, bamboo cabinet, wooden tables and chairs, grey tile and brick wall. The leaf shape wooden hollow partition and bamboo curtain as well as rectangle and polygonal window lattice cells has separated the space orderly. The combination of stone carving pillars and bamboo rails is chic, with the hollow design connecting the space of upper and lower storey. Festive red bridal sedan chair fully reflects the cultural characteristics of traditional Chinese custom, while stone carvings, tile carvings and wooden carvings that can be seen everywhere adds to the space a classical appeal. A few peach blossom are dotted indoors, which well demonstrates the ancient poem "peach blossom and rosy face set each other off". In order to pursue a serene, natural, pure and refreshing style, the architect has created an artistic synthesis that harmonize with natural environment.

新農莊

朱家

【东北人】

Northeasterner Restaurant

地址：上海水城路555号瑞泰酒店2楼
电话：021-62330990

『其人，大疆，勇而谨厚』，《后汉书·东夷传》中如是描写东北人的性格特征，大平原般荡气回肠和滚烫热辣话语豪爽。东北的饮食文化、关东风情和民俗文化也在本案中体现得淋漓尽致。

门厅以大面积的深色石块作为背景，前置东北特色大酒缸与石桌椅，无不流露出东北人的豪迈不拘，又恰似那片广袤的黑土地，让人深切地体会到了特有的关东风情。然而粗中有细，在射灯的重点照射下，不难发觉石块上精心雕刻着东北人的辛劳。设计师将中原文化融入东北风情中，在大面积的深色中点缀着点点星星的中国红，与之呼应的是中国明朝特色的灯挂椅，如此体现出东北文化的包容性。但是并没有本末倒置，青砖墙作为半墙隔断恰好分割出每个小单元。包间里舍弃了坐椅，垫高了地面，采用了东北传统就餐习俗——炕。在圆桌的就餐区域，设计师施以紫红色和花格，加上中式大宅门的红色，刺激着顾客的每一个味蕾，青砖墙与炕既给了顾客相对自由的空间，又让顾客感受到浓郁的东北民俗文化。本案自始至终都以人们耳熟能详的《东北人都是活雷锋》的歌词作为线索，穿插在餐厅的各个角落，又以东北特色的大红花布作为窗帘隔断，从细微处凸显出本案的主题『东北人』。

柳樹屯

翠花家

鱼籽酱
菜团子

The Book of the Later Han has described northeasterners as "strong, courageous, prudent and sincere" in Dongyi Chuan. Their character is soul-stirring and straightforward. Here the northeastern food culture, local customs as well as ethnic culture are fully expressed in this project.

The entrance hall is designed with northeastern style capacious wine vet as well as stone table and chair standing against extensive dark color stone background, revealing the bold and generous character of northeasterners just like the vast black land, and giving a unique northeastern flavor. Yet there are something refined contained in the bold. Under the spot light, you can find delicate carvings on stones telling the hard work of northeasterners. In the restaurant, the architect tries to infuse central plain culture into the northeastern flavor by dotting Chinese red in the dark background, which is echoed by Ming Dynasty lamp hang chair so as to demonstrate the inclusiveness of northeastern culture without putting the incidental before the fundamental. As a half wall partition, the black brick wall exactly separated every little unit. There is no chair in private rooms; instead they elevated the floor to adopt traditional northeastern dinning habit—heatable brick bed. The dinning area around the round table is designed carmine and lattice, which entices the palates when matching with Chinese red door. The black brick wall and the brick bed bring a relative free space and a strong northeastern flavor to guests at the same time. Throughout the space, the lyric of the famous song The Northeasterners Are All Living Leifeng is used as a thread that appears in every corner of the restaurant. Besides, the curtain is made of characteristic northeastern bright red cloth that highlights the motif of this project, northeasterner.

山裡屯

王老爷灶台
鱼
嘎

关东情

【西贝莜面村】

Xibei Restaurant

地址：上海浦东新区张杨路3611弄金桥国际商业广场1座4楼（金桥路口）

电话：021-58752999

西贝餐饮以西北民间菜肴为特色，凸显西北地域的饮食文化。设计师从打造品牌文化角度出发，拿捏有度且巧妙地在空间中选用体现地域文化特色的元素，营造出浓烈的西北民族风情与民俗文化。初入餐厅，简洁明快的几何吊顶造型，黑白红为基调的家具造型与局部软装，一种张弛有度的空间感受扑面而来。红黄色的空间色彩基调营造出热烈喜庆的氛围，让人情绪高涨，仿佛又被拉回到那个纯真质朴的年代。皮影戏中的人物剪影，白色的原木餐椅，民族乐器与戏曲人物结合的灯具造型，更是让人感受到浓浓的中国民俗文化。在浓厚的文化背景下用现代的形式展现出独具匠心的一面。墙面装饰的布老虎，手工绘制的腰鼓及传统的木质鱼造型，无不显示出西北人心灵手巧、热情豪放的性格特点。局部选用的中国红，热情而奔放，浑然而大气，除了饱含浓烈的背景文化外，更为贴切地凸显所营的菜肴特色，也展现出设计师在打造文化与空间形象融合的设计思路上颇为用心。

男洗手间
men's room

Featuring northwestern cuisine, Xibei restaurant highlights food culture of northwestern area. From the aspect of creating brand culture, the architect tactfully chooses elements that represent local culture to construct a strong northwestern flavor atmosphere in the space.

Upon arrival, a relaxed feeling comes to us. The interior is designed with the concise geometry form suspended ceiling, furniture and soft décor with fundamental tone of black, white and red. While the red and yellow tones create a jubilant ambience that keeps one's mood high as if taken back to those innocent years. The sketch of figures in shadow play, white log dinning chairs, ethnic music instrument and lanterns with traditional opera figures further extends a strong feeling of Chinese folk culture, showing ingenuity in a modern way in a profound cultural background. The cloth tiger, handmade waist drum as well as traditional wooden fish sculpt decorated on the wall is all demonstrating the intelligence and enthusiasm of northwesterner. The passionate and elegant Chinese red is applied in partial area to appropriately highlight the characteristic of the cuisine, demonstrating the architect's elaboration of blending culture into the image of this space.

手间
's room

【9车间川香工坊】

No.9 Workshop Sichuan Restaurant

地址：上海天钥桥路666号八万人体育场2号通道旁
电话：021-64266767

『9车间川香工坊』是车间还是餐厅？本案的主题就是车间文化，改革开放后的今天，我们忘却了粮票，也淡忘了最为熟悉的车间和大锅饭。置身其中，仿佛来到了20世纪五六十年代甚至更早的年代。本案的创意就在这里，一进门不见收银台，取而代之的是『传达室』，一旁立着一尊领袖雕像，墙上有：『自己动手，丰衣足食』标语。让顾客似乎穿越时空回到那个年代。餐厅装饰非常简单，灰水泥地面，灰格子钢窗，桌椅清一色的黑钢管，雪白的桌布，灰白两色，整个餐厅里置放着车床和老设备，看似简陋实则是用心设计。吊顶也未作大面积的改动，有意将屋顶结构暴露在外，打造一种形式美、结构美，同时舍弃了奢华的吊灯，用简单的灯泡与灯罩彰显着车间的特色文化。本案完全保留原车间结构，层高也得到了合理的利用，用黑钢管搭建起来的二层，依旧延续着车间的主题。墙上悬挂着的是那个年代的标语及海报，父辈来此家餐厅用餐，保证会引起他们很强的共鸣，从而讲述一段段他们当初的光荣岁月和人生感悟。作为9车间川香工坊，本案更加重于表现车间文化，但是也没有忽略『川香』的味觉信息，地面上墙面上用点点红色提起顾客『火辣』的热情。

命促生产
9車间
传达室
9車间
安全生产
门卫室
9車间
抓革命

9車间

灭火器箱
FIRE EXTINGUISHER BOX
火 119 警

安全出口
EXIT
传达室
9車间
9車间

全世界妇女团结起来 保卫世界和平
9車间
必须戴安全帽
9車间

就是
独立自主 自力更生
团
21

Is it a workshop or a restaurant? The motif of this project is workshop culture. At this era after reform and opening up, we have forgotten such a thing as food coupon and the once familiar workshop and communal pot also faded out of our memory. Entering this restaurant is like stepping into the 1950s and 60s, or even an earlier era.

Here lies the creativity of this project. At the entrance, a reception office took the place where usually a cashier stands. There is a sculpture of Chairman Mao beside the dinning tables, with the slogan, "Let's produce ample food and clothing with our own hands" on the wall, as if one has travelled across time to the last century. The decoration is very simple with grey cement floor, grey steel lattice window. The tables and chairs are made of black steel tube and covered by white table cloth. The restaurant is furnished with lathe and old equipment, crude yet delicately designed. The ceiling is barely modified to deliberately expose the roof structure, creating a beauty of structure. Meanwhile luxurious pendant lamps are not seen here; instead, the simple combination of bulb and lampshade demonstrates the feature of factory. The architect preserves the original structure of the workshop, and properly utilizes its storey height. The second floor build of black steel tube continues the motif of workshop by hanging slogans and posters of that era, which will definitely resonate with the elder generation who tend to tell stories of their good old days and perception of life. The project intends to present workshop culture, though; it does not forget to extend the taste sense of "Sichuan flavor", which finds an expression in the red dots on floor and wall finish to arouse the hot spicy passion of guests.

計划科

挽起袖子大干

9車间
川香工坊
25

9車间
川香工坊
28
29

【尊 悦】

Royaland Restaurant

地址：安徽宣城府山广场
电话：0563-2829111

从经营场所的规模和体量来看，这是一个大型的高档餐饮场所。设计师在设计上就必须把握好空间的尺度和奢华度，必须抓住商业经营者想要的尊贵感觉，并带给在此消费的顾客一种尊贵愉悦的心理感受。于是，设计师充分利用空间高度和广度。在门厅的处理上尽量表现出了一种大气豪华的感觉。拾级而上，在休息等待处，设计师将收银台设计成元宝形状，顶面倒挂的酒杯形状的灯具，即刻带来一种觥筹交错、杯光酒影、金碧辉煌的感觉，刺激顾客产生消费愿望，让经营者精神抖擞。沿着水族区，是一字排开的书法装饰墙，这种阵列式的处理手法，带来一种豪迈的气势和视觉冲击。包厢里，厚实的地毯和金碧辉煌的装饰让空间豪华的感觉油然而生，进而刺激消费。整个设计饱满充实，设计师在有充分空间尺度的设计舞台上长袖善舞，为经营者搭建了一个很好的经营平台。

须知
物各有先天，如人各有资禀。
性下愚，虽孔、孟教之，无益也。
物性不良，虽易牙烹之，亦无味也。指其大略：

老虎斑
石斑鱼
大扇贝
毛蛤
血蛤

In terms of the operation scale and size, this place is a large-scale high-end restaurant. Therefore the architect is required to have a perfect control upon spatial scale and level of luxury in design so as to convey to guests an honorable and pleasant feeling. As a result, the architect makes full use of the height and width of this space. The entrance hall is designed with a sense of grandness and luxury. Walking up the stairs, the waiting area consists of ingot shaped cashier and inverted glassed shaped lights hanging from ceiling, creating a resplendent image of wines, parties and toasts, which initiates guests' consumption desire. Along the aquarium area is a line of calligraphy wall. This array structure brings a vigorous momentum and visual impact. The private rooms are decorated with thick carpets and magnificent decors that naturally infuse the space with a sense of luxury and further stimulate consumption. By presenting this substantial and abundant design, the architect has built a great stage for the business operation.

一要用火先武后文，闷起得直。
放水，不多不少，燥湿得宜。
人家，讲菜不讲饭。逐末忘本
余不喜汤浇饭，恶失饭之本味故
佳，宁一口吃汤，一口吃饭，分前
方两全其美。不得已，则用茶
之，犹不夺饭之正味。饭之甘，
上；知味者，遇好饭不必用菜

【正院上海公馆】

Grand Mansion Cuisine

地址：上海浦东新区商城路665号C栋2-4楼

电话：021-58781888

正院上海公馆坐落于一幢美式官邸建筑内，经历了历史沉浮的老房子，到处弥漫着岁月的痕迹。设计师运用现代时尚元素及色彩，将中、西式设计风格结合得恰到好处。中国宫廷建筑的精华在现代美学元素的装点衬托下透露出无限典雅，既突出了中国文化的厚重，也尽现了西方美学的特点与时尚。步入餐厅，菱形青砖背景衬托着圆形的盘龙浮雕，配以栩栩如生的抱鼓石和各类陈设，营造出厚重的历史韵味。整齐排列的印章以镜面衬底，稳重又不失时尚感，搭配通透的洋酒酒架，又突出了中西贯通、古今融合的设计理念。沿着回纹装饰的过道，推开一扇扇门钉纵横排列的黑色大门，各具特色的包厢映入眼帘。高耸的空间以圆形吊顶搭配枝形大吊灯，紫色餐桌椅和精致的花纹壁纸在典雅中透露出华贵的气息，仿佛步入了欧洲宫廷。复式结构的『银锭观山』厅最为与众不同，挑高的二层空间设置了供客人休息洽谈的区域。大面积红色实木地板，楼梯上的紫色地毯，橙黄色的拱顶造型，紫红色的沙发，还有中式壁画上青翠欲滴的绿色，浓烈的色彩在一个空间中碰撞，尽显无限奢华。

太液秋风

Grand Mansion Cuisine is situated in the building of an American style mansion. Having experienced the vicissitudes of the history, the building is filled with the traces of time.

The architect employs modern fashion elements and colors to combine Chinese and western design style to just right. Dotted by modern aesthetic elements, the essence of Chinese palace architecture reveals an infinite elegance that not only highlights the dignity of Chinese culture, but also expresses the character of western aesthetics. Inside the restaurant, circular dragon relief is matched with diamond black brick background and other furnishings to create a strong historical aroma. The orderly disposed stamps are displayed on a mirror substrate, giving a prudent and yet stylish feeling and again emphasizing the design concept of "a combination of Chinese and western elements and of ancient and modern elements" while arranged with transparent wine shelf. Along the fretwork corridor, behind the black doors with vertically and horizontally arrayed doornails are characteristic private rooms. The lofty space is designed with circular suspended ceiling and chandelier. Purple dining table and chairs and delicate decorative wallpaper discloses an elegant and sumptuous flavor like that of a European palace. Duplex Yin Din Guan Shan hall is the most distinctive room. Its second floor is set for rest and meeting with extensive red solid wood flooring, purple carpet on the stairs, orange-yellow vault, claret-colored couch and green Chinese wall painting. All these intense colors crash with each other in one single space, revealing its luxury to the greatest extent.

安全出口
EXIT

【豫上海】

Yu Shanghai Restaurant

地址：上海旧校场路69号悦宾楼3楼（近沉香阁路）

电话：021-63283886

豫上海是一间打造『中西合璧，古色古香』饮食特色的主题餐厅，其坐落于熙来攘往的豫园老城隍庙附近，移步阳台，便可观赏庙内景观，以及餐厅所在的宏伟而又古色古香的传统建筑。

步入豫上海，迎面就会看到一幅独具特色的美女画像，将中国水墨画的技艺与西洋油画的风格巧妙融合，搭配吊顶处凹凸有致的黑白几何造型和书法、油纸伞等传统元素，传递出中西合璧的主题定位。主体空间以栅格式装饰造型为依托，高低错落的竹子悬挂造型颇具民族风情，贯穿空间的九曲桥和船造型的小舞台不定时演奏古筝、二胡等民族器乐，传递出中式古典韵味。点线面的原始设计元素与黑白色调的选用，又赋予空间现代气息。卡座区色彩艳丽的玻璃点缀，大面积地运用在吊顶上，打破了单一的色调，也为空间增添了哥特式教堂般的神秘韵味。餐厅还设有屋顶花园，一排柔软的沙发、一道和煦的阳光，营造出舒适的空间氛围，在吊顶处用红色油纸伞呼应了主题。设计师通过风格对比的手法，使豫上海完美地融合了古代和现代东西方风格，形成强烈的视觉冲击力。

Yu Shanghai Restaurant features a food culture combining Chinese and western style with antique flavor at the same time. As it is situated near the bustling Town God's Temple, guests can enjoy from the terrace the sight of the temple as well as the magnificent traditional building which is home to the restaurant itself.

Upon arrival, guests are welcomed by a unique beauty portrait that ingeniously integrates the techniques of Chinese painting and western oil painting, extending the motif of "combining Chinese and western style" along with the traditional elements such as black and white geometric form on the ceiling, calligraphy and oil-paper umbrella. The principle part is supported by lattice cell decoration, with bamboos of different heights hanging from the ceiling exuding an ethnic aroma. At times, there are folk music performance on the boat shape stage and zigzag bridge throughout the whole space, transmitting a Chinese classical aroma. While the original design elements of dot, line and plane as well as the selection of black and while color endows the space with modern flavor. In booth area, colorful glass ornaments are used in large area of the ceiling, breaking the monotonous tone and adding a mysterious gothic appeal. The restaurant has also a roof garden with a row of soft couches and a ray of vernal sunshine, which builds a comfortable atmosphere, while using red oil-paper umbrella on the ceiling to respond to the theme. Through a contrast of different styles, the architect perfectly mashes together classical and modern as well as Chinese and western, to form a strong visual impact. The restaurant not only represents tradition but also relevantly presents a rapid transformation of the old city.

出口
EXIT

【上海滩】

Shanghai Tan Restaurant

地址：上海黄陂南路333号企业天地2−3楼（近太仓路）

电话：021−63773333

说到上海滩，她繁华的灯红酒绿依然还留在一辈人的记忆里未曾褪色，依托浩瀚悠久的中华历史文化底蕴，融合外来多元文化，兼收并蓄，成就了她独有的魅力，吸引着一代又一代人对她的眷恋和神往。本案设计师正是在这样的定位下，试图再现上海滩那份独有的繁华魅力。首先从文化依托的角度，空间使用了大量的中式元素，比如八边形的窗户、苏州园林的格栅、冰裂纹的隔断，墙面彩绘也用了较多的梅花、云纹，使空间中飘荡着浓厚的中式意韵。上海滩的独特魅力不仅因她深厚的文化底蕴，更重要的是她对外来文化的兼收并蓄。传统的中式不论从造型或是用色上都较为保守和沉重，上海滩因特有的地理位置和历史进程，受到了较多的外来影响，为她的魅力增添了更多的表达方式，也冲淡了传统中式的沉重。比如本案的用色就十分大胆，大红大紫大绿在设计师的精心编排下不仅不显得凌乱，反而把上海滩的灯红酒绿表现得淋漓尽致。另一方面，鲜艳的色彩还能刺激消费者的感官，增强消费者的食欲，符合了餐厅的空间属性。这些大胆的用色，如行云流水般交织在中式元素之间，很好地起到了调和中式沉重的作用。

SHANGHAI TANG

Speaking of Shanghai bund, her prosperous image remains in the memories of that generation and has never faded. Relying on the long standing Chinese historical culture and blending it together with foreign multiculture, Shanghai bund has formed her unique charm that fascinates one after another generation.

Right under this positioning, the architect tires to recur the distinctive prosperous charm of old Shanghai bund. First, from the aspect of culture, the space still adopts a great number of Chinese elements, such as octagonal window, grille of Suzhou garden, cracked ice pattern partition. The wall painting is designed with plum and cloud patterns that infused into the space a rich Chinese aroma. The unique enchantment of Shanghai bund lies not only in her profound cultural deposits but more importantly in her inclusive spirit. Traditional Chinese style tends to be conservative and heavy in the selection of form and color. However because of the geographical and historical factors, Shanghai has received extensive foreign influence which enriches expression of her charm and offset the heaviness of traditional Chinese decoration. Take the use of bold colors for example, elaborately arranged bright red, purple and green colors, do not appear messy at all, instead, they have incisively and vividly described the feasting and revelry of Shanghai. Furthermore, bright colors can stimulate the senses of guests to whet the appetite, which corresponds to the nature of the space, namely restaurant. Those colors are naturally and smoothly interwoven in the Chinese elements, to offset the heaviness of Chinese elements.

HANGHAI TANG
Café

【上海，上海】

Shanghai Shanghai Restaurant

地址：上海嘉善路508号
电话：021-54655782

一提到上海，人的脑海中呈现的多半是电视剧里十里洋场的光怪陆离。抓住这样的第一感觉，设计师首先想到的应该是用色彩去诠释这样的上海故事。

这是个有着上下夹层的餐饮空间，在下部的空间，设计师着力营造着上海老宅子的设计概念。打开的窗扇，雪白的欧式风格椅子、七彩的玻璃门、优哉游哉的鸟笼子和四壁墙上的老式招贴画，让人的感觉在进门的一刹那慢下来，仿佛走进旧上海的风月生活。上到二楼，绿色成为装饰的主题，那么艳丽的绿，不论搭配什么都浮躁起来。在古典灯饰的映衬下，十里洋场的靡靡之音飘然响起。包厢里黑色的欧式椅子和墙面手绘的丝绸墙纸，默默无语中将奢华的感觉慢慢展现。走道上闪亮的灯具，暗淡的灯光，隐约中能听到古旧的留声机传出的舞曲，让人回到旧上海的欢愉中去，体味那个时代的文化，感受那充满烟草味的浮躁和属于那个时代有钱人的快乐。

SHANGHAI SHANGHAI
SHANGHAI
SHANGHAI
SHANGHAISHANGHAI

SHANGHAI SHANGHAI SHANGHAI SHANGHAI SHANGHAI

More often than not, the first that comes to one's mind at the mention of Shanghai is the fanciful old Shanghai in TV plays. Capturing this image, the architect intends to illustrate such a Shanghai story with colors. According to the drawing, it is a dinning space with double deck. At the lower level, the architect creates an old Shanghai house style space. Open windows, white European style chairs, colorful glass doors, leisure bird cage and old pictorial posters on the walls has compromised a view that slowed down the pace of time and lead guest into the romantic life of Old Shanghai. The second floor is featured with green, which brings a flashy look with whatever other decorations. Decadent music of old Shanghai is floating in the air against the classical lamp lighting. Black European style chairs and hand painted silk wallpaper are quietly revealing the luxury. Dance music from old phonograph can be heard faintly under the dim light at the aisle, enabling guests to feel the joy and culture of those days in old Shanghai among impetuous tobacco.

上海餅干

老熟地
老熟地
SILVERTON NORTHERN

【新上海】

New Shanghai Restaurant

地址：上海徐家汇路618号日月光广场4楼F19室（瑞金二路口）

电话：021-6093819

新上海主题餐厅，整体空间设计定位于一种用现代装饰手法讲述旧上海故事的设计思路，将老上海的万种风情通过一道道精致的美味和店面装饰展现到了极致。

小小的店面，外面看着没多大特别，而步入餐厅的一瞬间，便会被那股浓浓的怀旧气息所吸引。青砖墙与青石地面烘托出老上海的石库门风光，入口处风情万种的月份牌使造型简洁的服务台也充满了国色天香。改造后的灯挂椅，墙上的木质窗格，以中国红搭配原木色，古朴中又不失现代的气息。传统门扣装饰的包房大门，以老式的木质肌理展示着意味深长的历史风情。吊顶处的大面积黑色与原木横梁的选用，进一步强调了空间的主题，体现一种独特的人文气质。推开包房大门，大红色的墙面映入眼帘，灰色的木质桌椅置于中间，浓烈的色彩在灯光下映射出耀眼的光芒，相较于轻松惬意的大厅，又增添了华贵的气氛，让人感受到一种中国式的热情。设计师在入口墙面上设置了一处不规则的清玻，呼应了另一处断裂的墙面，使充满生机的鲜绿植物成为极好的框景，同时又将内部的景色以若有若无的姿态呈现出来，吸引外部人群的眼球，看似不经意的处理实则独具匠心。

NEW SHANGHAI
新上海

31
17

New Shanghai themed restaurant is defined as a space telling stories of old Shanghai by using modern decorative techniques. Its delicacies and decorations bring out the amorous feelings of old Shanghai to the extreme.

The restaurant has an ordinary façade; however, the moment you step inside, you will be attracted by its strong scent of nostalgia. Black brick wall and bluestone floor set off the old Shanghai stone gate and charming calendar picture adorns the simple service counter at the entrance. Transformed Ming style dinning chairs and wood windowpane are designed in Chinese red and log color, pristine and yet of modern flavor. Traditional door holders with vintage wood texture exude a feeling of historical lingering charm. The extensive black on the ceiling and log beam furthermore highlight the motif of the space, creating a unique cultural atmosphere. the private rooms are furnished with red wall surface and wood grey tables and chairs at the center. Strong colors shine upon the lighting and infuses a sumptuous feeling different from the relaxed hall, presenting to guests a Chinese style enthusiasm. The architect set an irregular shape glass on the wall of the entrance, which echoes to another cracked wall finish, making those vibrant green plants perfect enframed scenery as well as exhibiting the internal scenes faintly to draw attention from the outside. Every seemingly casual treatment of this space is actually created out of ngenuity.

23
22

图书在版编目（CIP）数据

中式餐厅/上海圣辉制版电脑有限公司编．福州：福建科学技术出版社，2013.2
ISBN 978-7-5335-4149-1

Ⅰ．①中… Ⅱ．①上… Ⅲ．①餐馆－室内装饰设计－图集 Ⅳ．①TU247.3-64

中国版本图书馆CIP数据核字（2012）第248475号

书　名　中式餐厅
编　者　上海圣辉制版电脑有限公司
出版发行　海峡出版发行集团
　　　　福建科学技术出版社
社　址　福州市东水路76号（邮编350001）
网　址　www.fjstp.com
经　销　福建新华发行（集团）有限责任公司
印　刷　深圳中华商务联合印刷有限公司
开　本　635毫米×965毫米　1/8
印　张　40
图　文　320码
版　次　2013年2月第1版
印　次　2013年2月第1次印刷
书　号　ISBN 978-7-5335-4149-1
定　价　320.00元